Pioneering Prosthetics
Advancements in Bionic Limbs and Exoskeletons

Table of Contents

Chapter 1. Introduction

In this special report, we delve into the fascinating world of "Pioneering Prosthetics: Advancements in Bionic Limbs and Exoskeletons". This compelling narrative navigates the intricate universe of groundbreaking medical and technological advancements in a manner suitable for both experts and novices alike. Unravel the once unimaginable world of bionic prosthetics and robotic exoskeletons, exploring how they have revolutionized the lives of countless individuals across the globe, granting not just renewed mobility, but fresh hope. This easy-to-understand yet comprehensive report provides a panoramic view of the field, extending from the basic principles of design to the transformative potential of these advancements for the future. Let's embark together on this riveting journey into the future of human potential – a future that is here today.

Chapter 2. The Evolution of Prosthetics: From Peg Legs to Bionics

The remarkable journey of prosthetics, charting its progression from wooden pegs and carved stones to the grip-responsive cyborg limbs of today, is a testament to human resilience, innovation, and a desire to improve quality of life for amputees. Let's delve into the fascinating ebb and flow of this evolutionary story.

2.1. The Dawn of Prosthetics

The necessity of prosthetics predates recorded history. The earliest known prosthetic, discovered in Egypt and dubbed the "Cairo Toe," dates back to between 950 to 710 B.C. Carved from wood, it was attached to the foot of its owner, indicating the quest for mobility and functionality started early in human consciousness. Around the same time, warriors and knights made use of crude wooden legs and iron hands during the crusades, signaling another phase in the early evolution of prosthetics.

2.2. The Middle Ages and Renaissance: Prosthetics for Functionality

During the Middle Ages and Renaissance, global conflicts sparked innovations in prosthetics with a focus on functionality. Famous examples include the iron hand designed for German knight Götz von Berlichingen in the 16th century. The hand could be manipulated into various positions, allowing him to grasp weapons or reins, marking a departure from the merely aesthetic prosthetics often used during this era. With the dawn of better tools and manufacturing processes, more elaborate and diverse prosthetics were created, elevating their importance within society.

2.3. The Surge of Innovation in the 19th Century

The 19th century saw an exponential surge in prosthetic design. Driven by the Industrial Revolution and a series of wars, prosthetics were improved with articulated joints, leather sockets, and even inclusion of simple mechanical functions. One example was the introduction of hinged prosthetics, which provided more natural movement and increased functionality. Soldiers, whose lives were irrevocably altered during wars, became the primary beneficiaries of these advancements enabling them to re-integrate into society with improved mobility.

2.4. Late 20th Century: From Passive to Responsive

The late 20th century marked an era of unprecedented advancements. Prosthetics moved from passive to responsive. Technological advancements allowed for the incorporation of lightweight materials such as advanced polymers and carbon fiber. Myoelectric prosthetics, marked by electrical activation of muscles, emerged. This era presented the 'Boston Arm', a prosthetic device that utilized residual electric signals from the user's musculature to produce subtle movements. This evolution brought about a significant upgrade in terms of both functionality and aesthetics for users of prosthetic limbs.

2.5. The Birth of Bionics: The 21st Century and Beyond

As we moved into the 21st century, the line between man and machine began to blur, with bionic prosthetics ushering in a new era. Bionic limbs controlled by the user's mind offer unprecedented levels of functionality. They come equipped with AI systems that learn the user's pattern of movement, and sensors that provide haptic feedback of different textures. Cutting edge technologies such as 3D printing have become invaluable, reducing costs and enabling customization, bringing bionic prosthetics within the reach of many who need them.

2.6. Challenges, Future Prospects and Final Thoughts

As impeccable as these advancements seem, there remain numerous challenges. The integration of bionics into the human body still poses significant hurdles, including rejection by the immune system, energy efficiency issues, and high development costs. However, armed with the power of technology and an ever-evolving understanding of human physiology, scientists and engineers are working to overcome these challenges.

As we continue to unlock the marvels of neuroprosthetics and nanotechnology, the future holds untold potential for the evolution of prosthetics. Imagine a future where advanced bionics not only mimic but also exceed human capabilities, perhaps allowing us to experience the world in entirely new ways. We may not quite be there yet, but the boundary-pushing field of prosthetics signals that such a future may not be far off.

This evolution, from peg legs to bionics, is a stark testament to human innovation and our determination to ensure quality of life, irrespective of physical limitations. Remaining at the forefront of technology, the world of prosthetics continues to redefine the essence of human potential, offering a beacon of hope and a masterpiece of modern medical engineering.

Chapter 3. The Anatomy of a Bionic Limb

The metamorphosis of prosthetics from crude, rudimentary extensions to highly sophisticated bionic limbs has been anything but ordinary. Bionic limbs, often termed as the epitome of modern technology incorporated within the human body, promise to redefine the very core notions of disability and impairment.

3.1. Constructing The Bionic Limb

Every bionic limb consists of certain fundamental elements that work together to mimic the natural movement of a human limb as closely as possible.

Let's take a deeper look into these integral components:

- Controller: This element orchestrates the overall functioning of a bionic limb. It follows a series of commands relying on algorithms that are programmed to interpret electrical signals from the body of the wearer. Essentially, the work of the controller can be likened to that of a computer's CPU.

- Sensors: These invaluable components pick up the signals from the wearer's residual limb or another part of the body. The sensors may detect muscle contractions, brain activity, or even heart rate to provide inputs to the controller.

- Actuators: Actuators are the mechanical elements that function like human muscles. They execute the actions initiated by the controller, enabling the limb to move, grip, or walk.

- Power Source: To ensure the smooth running of bionic limbs, a reliable and long-lasting power source is the need of the hour. This is usually provided in the form of rechargeable batteries.

3.2. Harnessing The Power of Thought

One of the critical traits that make bionic limbs revolutionary is the ability for the wearer to control the limb through thoughts alone. A process known as 'Targeted Muscle Reinnervation' (TMR) makes this possible.

In TMR, surgeons redirect the nerves responsible for controlling the muscles in the lost limb to the remaining muscles in the residual limb. When the amputee thinks about flexing the lost limb, the brain dispatches the commands to the redirected nerves, causing a visible contraction in the muscles of the residual limb. These muscle movements are then detected by the sensors in the prosthetic limb, thereby initiating movement in the bionic limb.

3.3. Key Features of Bionic Limbs

- Improved versatility: Bionic limbs are technologically equipped to perform a broader range of movements compared to traditional prosthetics. This could include complex movements such as opening and closing of the hand, flexing the wrist, or even twisting an ankle.

- Enhanced Force Sensing: Modern bionic limbs come embedded with force sensors within the fingers, allowing the users to have a better understanding of the grip-force utilised. This equates to a better hold on objects, reducing the chances of slips or drops.

- Flexibility and Customisation: Bionic limbs offer the possibility of adjusting the speed, force or grip patterns to better suit the comfort and needs of the individual. This level of customization provides a more user-friendly and natural interface for the wearers.

- Improved Communication: Some versions of bionic limbs can

communicate with smartphones via Bluetooth technology. This capability carries a potential for the advancement of self-diagnosis or long-distance troubleshooting, making it easier for wearers to manage and maintain their bionic extensions.

3.4. No Future Without Challenges

While the advancements in bionic prosthetics hold considerable promise for the future, there also exist several challenges that could potentially hinder further progress.

From the economic standpoint, the high cost of designing, manufacturing, and maintaining bionic limbs makes it inaccessible for a majority of the population. Moreover, the need for regular adjustments, recalibrations, and in certain cases, replacements, only adds to the problem.

On the technical front, issues like weight, noise, and the requirement of a reliable power source are glaring drawbacks that need immediate attention.

However, despite these challenges, the field of bionic limbs holds considerable optimism. Each challenge faced paves the way for a multitude of research opportunities, marking the field as an exciting sphere of constant development and innovation.

3.5. Gazing Into the Future

The field of bionic limbs is continually pushing boundaries to create solutions that are less distinguishable from natural limbs, both in terms of aesthetics and functionality. With ongoing development in machine learning and artificial intelligence, the dream of a 'natural-feeling' bionic limb is inching ever closer to reality.

This journey into the anatomy of bionic limbs paints a vivid picture of the exceptional leaps we have made in prosthetics. As we continue to integrate the fields of neuroscience, robotics, and bioengineering, the horizon of human potential is poised to witness an era where disability ceases to hinder mobility, thus breathing fresh hope into the spirits of countless individuals across the globe.

Chapter 4. Exoskeletons: Beyond Fiction towards Reality

4.1. Introduction

The concept of exoskeletons, traditionally confined to the pages of science fiction novels and the silver screen, has rapidly evolved into a tangible reality that is already positively impacting the lives of many. These are essentially wearable structures that work in sync with the human body, leveraging the principles of mechanics, electronics, sensors, and software to augment the body's existing physical capacity, and in some cases, provide mobility where none existed.

4.2. Historical Development

The concept of an exoskeleton is not as modern as it might seem. In fact, the idea dates back centuries, to the days when Leonard da Vinci sketched out plans for a robotic knight. Despite da Vinci envisioning this technology long ago, the first truly functional exoskeleton project, named HARDIMAN, wasn't started until the late 1960s by General Electric in partnership with the U.S. Military. However, the technology of this time was not advanced enough to produce the desired results, and the project was deemed unsuccessful.

Fast forward to the 21st century, the advancements in technology have paved the way for developing exoskeletons that can augment human strength, aid people with mobility issues, and protect workers from injury.

4.3. Design and Functionality

The primary purpose of exoskeleton design is to create a wearable device that seamlessly complements and amplifies the human body's capacities. It involves a blend of mechanical design, control systems, sensors, and human-machine interfaces. The design, size and weight of the exoskeleton are critical factors influencing the user's comfort and acceptance.

To achieve this, the exoskeleton employs a range of sensors to monitor movements and provide real-time feedback to the user. Actuators mimic the muscle's role, while a control system understands the intended actions and directs the actuators accordingly.

4.4. Classification of Exoskeletons

Exoskeleton designs can be broadly classified into two categories: passive and active.

- Passive Exoskeletons: Passive exoskeletons are devices that do not utilize any form of power supply for their operation. They primarily rely on mechanical elements like springs and dampers, helping redistribute the body's weight and reduce muscular strain.

- Active Exoskeletons: Active exoskeletons are equipped with actuators that provide assistive force to the wearer. These systems are often electrically powered and utilize sensors and complex algorithms to sync with the user's actions.

Both types carry unique strengths, suiting different types of tasks and user requirements.

4.5. Exoskeletons in Rehabilitation

One of the most impactful uses of exoskeleton technology is found in the field of rehabilitative medicine. Stroke victims, spinal cord injury patients, and individuals with mobility impairing conditions have been able to regain movement and independence via these revolutionary devices.

In physical therapy, exoskeletons not only encourage movement for patients but also provide biofeedback for clinicians. This helps physicians monitor and alter therapy, leading to more effective results. For people with impairments, the technology has showcased incredible potential in restoring their ability to walk, hold, or move again, offering a profound shift in their quality of life.

4.6. Exoskeletons in Industry

Exoskeleton technology is expanding beyond the medical arena and making waves in Industry. Industries worldwide are starting to recognize the perennial value of exoskeletons in enhancing productivity while ensuring worker safety.

In construction and manufacturing, exoskeletons help reduce the physical load on workers, saving them from musculoskeletal disorders. In the military field, exoskeletons can give soldiers superhuman strength, protect them from physical harm, and significantly improve their endurance.

4.7. The Future of Exoskeletons

The potential of exoskeleton technology goes well beyond its current applications. Even as you read this report, innovators across the globe are pushing past the boundaries of what exoskeletons can do.

We are already witnessing the integration of AI and machine

learning into exoskeleton design, enhancing their adaptability and capability. The amalgamation of virtual and augmented reality technologies presents another exciting avenue.

That said, the development of exoskeleton technology isn't without challenges. The issues of affordability, accessibility, regulatory compliance, and user acceptance have to be carefully addressed, ensuring these devices can benefit those who need them most.

The future beckons a time where the delineation between humanity and technology blurs increasingly, allowing us to explore the possibilities of our evolution. With the ever-improving gears of innovation turning steadily, we look forward to the transformative promises the future of exoskeleton technology brings.

Adventures, once rendered constrained by biological limitations, breathe anew in the canvas of possibilities that exoskeletons unfold. The quest for human augmentation has entered a fascinating chapter, painting a compelling picture of our collective future, filled with the promise of fostering a more inclusive, productive, and boundary-less society. Behind the scenes, the dance of disciplines ranging from medicine, engineering, AI to psychology continues to enrich this remarkable symphony of progress. Science fiction truly doesn't remain fictional for long.

Chapter 5. Underlying Technologies Powering Prosthetics and Exoskeletons

Prosthetics and exoskeletons, once confined to the realm of science fiction, are reshaping the landscape of mobility and rehabilitative care in the modern era. The technological advancements underlying these devices - the software, the intricate hardware, and the human-machine interfaces - are nothing short of revolutionary. Walking, running, holding, and even high-precision tasks like writing or painting, once deemed impossible with artificial limbs, are now becoming a reality.

5.1. Breaking Down the Components

A robotic prosthesis or exoskeleton consists of multiple components working in harmony. There's the structural body, which physically resembles a human limb or torso, typically made from lightweight materials like carbon fiber. This body houses the powerhouse of the device: actuators, which mimic muscles and joints, controlling the movement.

In powered prosthetics and exoskeletons, these are typically electric motors or hydraulic systems. Motors are frequently used due to efficiency, noise levels, and power-to-weight ratios. An AI-driven control system commands these actuators, driven by inputs from sensors that measure variables like position, speed, force, or even electrical impulses from a user's muscles or nervous system.

5.2. Mechanics and Actuation

The heart of any prosthesis or exoskeleton is its actuation system.

Similar to how our muscles contract to produce movement, actuators power the artificial joints of these devices. Motors create rotation, which is converted into the necessary linear or rotational movement via a series of gears.

Advancements like brushless DC motors have improved efficiency, providing more power from smaller motors, while increased battery efficiency means devices can operate for longer periods. Hydraulic actuators, on the other hand, often provide higher forces and faster responses, but require more complex equipment and consume more energy.

5.3. Sensors and Feedback

Creating responsive and natural-feeling devices requires a robust feedback system. This is achieved using various sensors that detect forces, joint angles, accelerations, and more.

Force sensors might be placed to measure ground reaction forces, or within the device's "muscles," to register the forces they exert. Gyroscopes and accelerometers can provide information about the device's orientation and movement, critical for maintaining balance and avoiding falls.

Moreover, some cutting-edge prosthetics utilize biofeedback, directly using electrical signals from a user's nervous system to control the device. Myoelectric sensors detect these signals through the skin – the stronger the signal, the stronger the response from the device.

5.4. The Role of AI and Machine Learning

These advanced devices couldn't work without sophisticated algorithms interpreting signals from the myriad of sensors and adjusting the actuators accordingly. Machine learning algorithms use the user's inputs to optimize the device's responses over time, helping the machine become a natural extension of the user.

These systems need to be extraordinarily efficient due to the complexity of human movements and the vast number of sensory inputs to interpret.

5.5. Advanced Materials

Lightweight yet robust materials are vital for maximizing user comfort and maneuverability while also ensuring durability. Carbon-fiber composites are commonly used due to their strength-to-weight ratio and are becoming more cost-effective.

Similarly, advanced flexible materials like shape-memory alloys are being explored, which could enable prosthetics and exoskeletons to mimic the varying stiffness and flexibility of human muscles and joints.

5.6. Challenges and Future Directions

While these technologies have promised much, they are not without challenges. Ensuring compatibility between the human and the machine is one such concern. Biofeedback systems, for example, often require surgery to implant electrodes, limiting their accessibility.

Moreover, these devices need to be affordable and accessible for them to have broad societal impact. Innovations in 3D printing, including printing with conductive materials, could lower these barriers, enabling more widespread application of these technologies.

The future holds immense promise. We're already seeing the advent of brain-computer interfaces, where signals from the brain are translated directly into device control.

This whirlwind tour underscores the complexity of these technologies, their vast potential, and the appreciation we should have for our own biological systems. Technologies powering these devices are not just about improving the mechanics, but about understanding and harnessing the complexities of human bodies, the very essence of biomimicry.

Chapter 6. The Human-Machine Interface: Making the Connection

Understanding the symbiotic relationship between a human and a machine requires a deep dive into the complex world of the human-machine interface (HMI). This vital link allows for the seamless integration of bionic limbs and exoskeletons with the human nervous system, truly embodying the concept of "cybernetic organisms". In essence, HMI emphasizes the harmonious functioning of artificial limbs or devices with human intention. How, then, is this intricate connection forged and maintained? This chapter will articulate this question's nuances, unfolding the science and technology underlying human-machine interfacing.

6.1. The Magical Manifestations of Neuroplasticity

The astounding ability of the human brain to adapt is termed as neuroplasticity. This inherent neural flexibility is the bedrock upon which the foundation of effective HMIs is laid. When an individual loses a limb, the brain's areas serving that limb don't become 'silent'. Instead, these areas reconfigure themselves, maintaining activity levels and creating new pathways.

In the bionics realm, neuroplasticity enables the brain to accept and integrate artificial limbs, creating a bi-directional exchange of information. Scientists tailor sensory feedback mechanisms in prosthetics that can transmit signals to the brain, which then interprets these signals like it would from a biological limb.

Advanced techniques are also employed to tap into the neuroplastic properties of the brain. Targeted muscle reinnervation (TMR) and Selective Nerve Transfers are surgical procedures that re-route residual limb nerves to remaining muscles. This enables the person to control their prosthetic limbs merely by thinking about the movement they want to perform.

6.2. Bionic Touch: Achieving Sensory Feedback

Sensory feedback in bionic limbs replicates a vital function of biological limbs - the sense of touch. Employing mechanisms like force sensors and electrodes, artificial limbs are designed to detect and transmit sensory data back to the brain. These sensations can span from the pressure of a handshake to the texture of an object.

The use of tactile feedback systems within prosthetics allows users to understand the grip force exerted, enabling them to manipulate delicate items without damaging them. Thanks to the advancements in embeddable sensor technology and machine learning systems, the consistency and accuracy of this feedback have seen significant improvements.

6.3. The Innovations in Exoskeletons: Beyond Prosthetics

While prosthetics replace lost limbs, exoskeletons are designed to augment human physical abilities or make up for reduced functions, primarily in terms of mobility. In essence, these are wearable machines that exploit HMIs to respond to human intentions, mimicking, augmenting, or even enhancing body movements.

Exoskeletons are primarily used in rehabilitation therapy for stroke or spinal cord injury patients, assisting and encouraging motor functions. They are also used in industrial applications to support workers in performing strenuous tasks more efficiently, effectively reducing work-related injuries.

HMIs within exoskeletons utilize methods like electromyography (EMG) to pick up electrical signals from muscle contractions. Advanced algorithms then interpret these signals to initiate corresponding movements in the exoskeleton.

6.4. Non-invasive Vs Invasive Interfaces: Weighing the Pros and Cons

Human-machine interfacing in bionics can either be non-invasive, usually employing surface electrodes, or invasive, with implanted devices. While non-invasive methods are more common due to lower risks, they tend to have limited control capabilities and can be affected by external factors like sweat or muscle fatigue.

On the other hand, invasive methods, like implanted myoelectric sensors (IMES) or cortical implants, offer improved selectivity and signal quality. However, they come with the risks associated with surgical implantation and the challenge of device miniaturization.

6.5. The Path to Intuitive Control and Independence

Refinements in HMIs, backed by powerful computational algorithms and machine learning, strive to create a more intuitive control for bionic limbs and exoskeletons. The journey from operating buttons and switches to mind-controlled movements reflects significant advancements.

The ultimate objective remains to reduce dependence on external tuning and calibration. The incorporation of adaptive learning systems that understand user behavior over time is a promising step in this direction.

6.6. The Road Ahead: Challenges and Opportunities

Bionics has certainly come a long way, thanks to inter-disciplinary integration and technological advancements. However, certain bottlenecks persist, such as perfecting sensory feedback, reducing power consumption, and ensuring device longevity.

While integrating more sensory inputs can enhance functionality, it needs sophisticated processing. Moreover, designing a compact yet powerful power supply remains a challenge. Material science advancements point towards the development of body-compatible, durable materials for long-lasting devices.

The advancements in HMIs carry immense potential that extends beyond empowering individuals with physical limitations. In industries, where manual tasks can be hazardous, wearable exoskeletons can offer safer alternatives. In sports and entertainment, bionics could redefine human potentials. The fusion of science, technology, and creativity in the field of bionics assures a future that was once in the realms of science fiction, is now evolving into a reality.

Chapter 7. Real-World Applications of Bionic Limbs

If addressing the dynamic possibilities of a realm as evolving as prosthetics, it's imperative to bring into focus the real-world implementation of these prosthetics. With every engineering marvel, the most crucial hurdle to cross is its aptness to be employed in real-world situations. This comprehensive narrative captures the substance of real-world applications of bionic limbs—in an environment far removed from the confines of a research lab.

7.1. Medical Applications

Bionic limbs are often seen changing lives in the medical arena. Those born with limb deficiencies or have had limbs amputated due to accidents or disease can regain their mobility with advanced prosthetics. In the past, amputees were fitted with rudimentary prosthetic limbs that served the basic purpose of aesthetics with minimal functionality. Today, advanced bionic limbs provide the possibility for individuals to perform complex movements almost on par with natural limbs.

The case of Hugh Herr, a bionic limb user and researcher himself, serve as prominent examples. After losing both his lower limbs due to a climbing accident, Herr went on to design sophisticated bionic limbs that enabled him to return to his beloved sport. Today, his developments have assisted countless amputees regain their mobility and independence.

7.2. Sports and Recreation

Bionic limbs have transformed the arena of sports and recreation as well. Not only have they enabled individuals to resume their participation in a myriad of sports activities, but they have created new opportunities for competitions designed specifically for athletes using assistive technology.

Take, for example, the Cybathlon—an Olympics styled international competition for athletes with physical disabilities, where they compete using advanced assistive devices, including robotic prostheses. It creates a platform where cutting-edge advancements in prosthetic technologies can be demonstrated, contributing to a more inclusive global sporting arena.

7.3. Military Applications

Moving from the sporting arena, bionic technology has found profound implications in military applications. Rehabilitation for wounded soldiers has improved dramatically with the advent of advanced bionic limbs.

Programs such as the Defense Advanced Research Projects Agency (DARPA) in the United States invest heavily in these technologies for rehabilitation and to improve the quality of life of veterans. Advanced bionic limbs provide the possibility for wounded soldiers to return to active service or make their civilian life physically less restrictive.

7.4. Integration with Emerging Technologies

Emerging technologies such as virtual reality (VR), augmented reality (AR), and artificial intelligence (AI) have begun integrating with bionic limb technology to enhance user experience and future potential benefits.

Virtual and augmented reality have emerged as effective training tools for new prosthetic users. They can simulate real-world environments and challenges, helping users adapt to their new limbs before the physical world trials. AI also plays a crucial role as it forms the core of learning algorithms in bionic limbs, enabling them to learn and adapt to the individual gait patterns of different users. The integration of these technologies with prosthetics opens up a whole host of opportunities and potentials in the real-world application of bionic limbs.

7.5. The Road Ahead

The real-world applications of bionic limbs span an impressive range, from providing mobility to individuals, empowering sports personnel, aiding in military rehabilitation, to integrating with other emerging technologies. Despite these advancements, challenges like cost effectiveness and accessibility remain. However, with continued research and the pursuit of innovative solutions, the distance between what's currently possible and a future where high-quality bionic limbs are accessible to all, is progressively blurring.

Current advancements in technology have worked in favor of pioneering prosthetics, marking a monumental change not just in the mechanical aspect but in the societal response towards it. With varied real-world applications, bionic limbs are propelling the world into an age where losing a limb doesn't equate to losing mobility, securing a future that promises inclusion and new found independence.

The journey of bionic limbs from restricted lab settings to real-world applications is a story of tremendous achievement. It is an evolving tale of human resilience, scientific persistence, and the audacious belief in breaking boundaries—an adventure that ensures technology serves humanity in its deepest sense.

Chapter 8. Exoskeletons in Practice: Rehabilitation and Mobility

Technological advancements in exoskeleton design over recent years have remarkably enhanced the possibilities for the rehabilitation and mobility among individuals with physical limitations. In practice, exoskeletons serve as wearable machines driven by a system of motors, pneumatics, levers, or hydraulics, manipulated by the wearer to mimic or augment their mobility abilities.

8.1. The Basics of Exoskeleton Design

Understanding exoskeletons in practice hinges on acknowledging their basic principles of design and function. Each exoskeleton is custom-fit to the user to ensure effective operation. Traditional prosthetics replaced missing limbs but lacked the dynamic movement abilities inherent to exoskeletons. Today's modern exoskeletons mirror the complexities of the human body: they incorporate weight distribution, mobility, and balance features that acquire electrical potentials from the brain, muscles, or the user's intended motion, ultimately radiating the desired movement effect.

8.2. Rehabilitation Clinics: A Focal Point

One primary arena where exoskeletons have been fully ingrained into functional use is rehabilitation clinics. These institutions harness the power of robotic suits to enable patients, particularly those suffering from spinal cord injuries or strokes, to regain control of their limbs and relearn motions that were once second nature. In fact, the introduction of exoskeletons in these settings has massively eliminated the need for labor-intensive traditional physiotherapy and enhanced recovery timelines.

The EksoGT, for example, is used globally in clinical rehabilitation for individuals with various degrees of paralysis. Its controlled environment and real-time adjustments make it an effective tool for retraining the body and rebuilding neural pathways, elements vital for walking.

8.3. Daily Life Integration: Improving Mobility

Parallel to their use in rehabilitation, exoskeletal devices have also infiltrated the daily lives of many. People with paralysis or longstanding disabilities utilize different exoskeleton designs to perform daily activities effectively and independently.

One comprehensive example of this is the ReWalk system, an exoskeleton that allows paraplegic individuals to walk again. This device embraces user-initiated mobility; tilt sensors in the device respond to shifts in balance from the wearer, prompting the exoskeleton to take a step and allowing movement that meshes with the natural rhythm of the user.

8.4. The Military and Industrial Applications

The scope of exoskeleton use extends beyond medical and personal use; sectors such as the military and industry are also benefiting from these innovations. Several military initiatives worldwide explore exoskeletal suits in improving soldier endurance, speed, and lifting capabilities - the USA's TALOS (Tactical Assault Light Operator

Suit) being one notable example.

For industrial applications, the Bionic Power's PowerWalk aims to capture and uses energy generated during human locomotion for charging a variety of devices. Likewise, the suitX Modular Agile eXoskeleton (MAX) is designed to prevent heavy lifting injuries at worksites, by providing innovative solutions that minimize the stress on knees and the lower back.

8.5. Future Prospects: Paving the Road Ahead

The future of exoskeleton technology holds exciting possibilities: designs with enhanced sensory feedback capabilities, more lightweight models, and outspread use in augmenting human abilities in various fields. The rising trend of integration with AI suggests 'intelligent' exoskeletons - machines learning from the user over time, adapting to their patterns and requirements, turning them into personalized assistive devices.

In summary, exoskeletons mark a glorious reimagining not just in the rehabilitative medicine field but likewise, in a unique blend of advanced technology enhancing human life quality. These wearable machines have and still are tipping the scales from disability to immense possibilities, rewarding users with a newfound sense of freedom and control, ultimately reshaping societal perceptions of physical limitations and disability.

For both the experts and the novice, the exoskeleton represents an exciting shift in our perception of human potential – a perception that is being redefined, reassessed, and revolutionized constantly. The integration of exoskeletons into various aspects of human life opens doors, redefines possibility, and within that, redefines what it means to be human.

Chapter 9. Challenges and Limitations in the World of Advanced Prosthetics

Though advancements in bionic limbs and exoskeletons have garnered significant attention for their potential to revolutionize mobility and quality of life for those with physical impairments, they are not without their challenges and limitations. Below, we explore these issues extensively to provide a comprehensive understanding.

9.1. Technological Constraints

While the technology has been increasingly refined with time, there remain certain persistent technological constraints. First, powering these advanced devices efficiently and effectively continues to pose a problem. Prosthetics and exoskeletons often require highly specific, high-capacity power sources that are both portable and lightweight, considering their intended wearability. As of now, no solution has fully met these criteria, leaving room for more innovation in the field.

Secondly, the issue of system control in bionic limbs is another formidable challenge. To mimic the human body's natural movements, convoluted algorithms and precise motor control are needed. As a direct consequence, these devices become complex and might require extensive training to use, thus limiting their usability to some extent.

9.2. Physiological Considerations

In tandem with the technological hurdles, physiological constraints also significantly influence the development and implementation of these devices. To start with, there's the key issue of 'socket comfort.' An uncomfortable socket can lead to discomfort, skin irritation, and injuries over time, rendering the advanced prosthetics useless.

An additional physiological challenge arises from the lack of proprioception, or body awareness, that amputees often suffer from. People who have lost a limb often struggle to adapt to advanced prostheses due to their altered perception of their body's position and movements in space.

Integrated solutions are being investigated to help amputees re-train their brain to recognise the bionic limb as part of their body and therefore increase user acceptance and improve quality of life.

9.3. User Acceptance and Training

User acceptance of advanced limb prostheses and exoskeletons is an essential factor to consider. Technological advancement notwithstanding, if the device fails to achieve acceptance among users, its potential applications are largely negated. Much of this acceptance hinges on how well users can adapt to the prosthetic or exoskeleton and whether it feels natural to them. Unfortunately, due to the complex and often counter-intuitive control strategies employed, many users struggle with this adaptation process.

In addition, the substantial amount of training required to successfully operate these devices can also act as a deterrent to many potential users. Adequate training is key to the successful integration of the prosthesis into the individual's daily routine, and the lack of such training can significantly hamper user acceptance.

9.4. Cost and Accessibility

The high cost of bionic limbs and exoskeletons is another significant barrier to their wider adoption. Given their technologically advanced nature, these devices are often prohibitively expensive for most potential users. This limitation, combined with often costly and lengthy rehabilitation and training requirements, makes accessibility a sizable challenge in the realm of advanced prosthetics.

9.5. The Path Forward

Addressing these challenges will require a concerted effort from researchers, manufacturers, healthcare professionals, and policy makers. Solutions need to be found that balance advancements in technologies with the users' physiological and psychological well-being. Additionally, measures should be taken to reduce cost and improve accessibility, thereby ensuring that these revolutionary advancements are available to all who need them. As the technology continues to evolve, so too must our approaches to these challenges and limitations.

In conclusion, while the world of advanced prosthetics is filled with promise, it is simultaneously riddled with deep-seated challenges. Although these hurdles may appear daunting, they delineate the pathway for future advancements in the field. By persevering and persisting in our efforts to overcome these barriers, we can enrich and indeed transform the lives of countless individuals across the globe.

Chapter 10. Ethical and Social Implications of Bionic Technology

Technological advancement is often fraught with manifold ethical and social implications, and the field of bionic prosthetics and exoskeletons is no exception.

10.1. Ethical Considerations in the Development and Use of Bionic Technology

Like any healthcare technology, prosthetic limbs and exoskeletons exist within a complex ethical framework. Principally, clinicians, biomedical engineers, and policymakers juggle questions of fairness, accessibility, patient autonomy, and the potential for unforeseen consequences.

Perhaps the most heated discourse revolves around access and fairness. Advanced prosthetic limbs are high-cost medical solutions, with high-end models reaching upwards of $50,000. This significant financial barrier is prohibitive to many individuals, particularly those living in socioeconomically challenged regions or without comprehensive health insurance. The question then arises, should everyone not have equitable access to these life-changing technologies regardless of social standing or economic resources?

Moreover, even if equitable access can be achieved, there are other issues of fairness to consider. As these technologies continue to advance, the line between medical necessity and enhancement blurs. In the future, bionic limbs might surpass natural human ability, raising the question of whether there should exist a limit to how advanced prosthetics may be for non-therapeutic use. This leads into a broader discussion about the definition of disability and what it means to be human in the age of bionic technology.

Patient autonomy is equally crucial in these ethical deliberations. As bionic limbs become more integrated with the human body – linked to nerves and controlled by brain signals – they move beyond mere tools to parts of the person's self. Consequently, questions of bodily integrity, identity, and the right to control one's body become central to the narrative.

Lastly, there is the concern over the unforeseen consequences of the rapid advancement of bionic technology. Could these technologies be weaponized? Would they lead to a stratified society, where "bionic" individuals are looked upon as superior? These necessitate a proactive and dynamic regulatory framework that can adapt to evolving technologies.

10.2. Social Implications and Responses to Bionic Advancements

Societal responses to bionic prosthetics and exoskeletons have been overwhelmingly positive, as seen in their depiction in popular media as symbols of overcoming challenges and heightening human potential. However, there are areas of persistent social tension.

Acceptance and stigma remains an issue for recipients of bionic limbs and exoskeletons. New recipients may experience a range of reactions from those around them, ranging from admiration and respect to discomfort or fear. This is particularly evident when prosthetics are fully visible, drawing attention to the recipient's altered body.

Additionally, a lack of understanding about these technologies and their use can further contribute to discrimination or isolation for users. In some cases, recipients face misconceptions about their abilities – such as a presumed athletic advantage in sports competition. More needs to be done to educate society about these technologies, pushing past sensationalism to understanding.

Rapid advancements also pose challenges to both personal and cultural identity, as the integration of artificial parts changes perceptions of self and humanity. Is an individual with a bionic limb still entirely themselves, or partially a machine? However, many with prosthetics have found empowerment and positive identity reaffirmation through their new capabilities.

Finally, the rise of bionic technology leads to profound conversations about human augmentation and ability. Is there a point where enhancement becomes unfair? Should we allow athletes with bionic limbs to compete alongside those without? How do we ensure inclusivity and fairness in a rapidly evolving technological landscape?

10.3. Conclusion: Navigating the Future of Bionic Technology

As we proceed into a future increasingly marked by the presence of bionic technology, it is essential that we regularly reevaluate the ethical and social discussions surrounding it. Balancing the interests of individual users, society, and the potential future implications of these technologies is critical. The hard conversations, from access and fairness to personal identity and societal understanding, underpin the development and implementation of these life-changing tools.

Understanding bionic technology's social implications and confronting its ethical dilemmas will ensure that the science does not precede compassion and consideration. The ultimate aim should be to create a future where these technologies are used responsibly – empowering individuals, improving quality of life, but not at the cost of our shared humanity. Navigating this terrain will certainly be complex, but if history is any predictor, it will also be irresistibly fascinating.

Chapter 11. A Glimpse into the Future: Emerging Trends and Innovations

Emerging trends and innovations in the field of prosthetics and exoskeletons are charting the future of human possibilities. Advancements in technologies such as artificial intelligence (AI), robotics, and nanotechnology are bringing us closer to a world where limb loss and disability no longer hinder human potential.

11.1. The Power of AI and Machine Learning

AI and machine learning have become groundbreaking in the realm of bionic prosthetics and exoskeletons. Researchers are integrating these technologies into bionic limbs to make them more adaptable and in sync with the user's requirements. For example, machine learning algorithms are being used to recognize and predict the user's movements, increasing the functionality of the prosthetic device. This development significantly improves the usability and performance level of bionic limbs, putting them one step closer to mimicking the natural human movement.

In addition, AI involved in error correction and fine-tuning of movement ensures the continuous adaptation of the prosthetic limb to the user's needs. Rather than the user adapting to the prosthetic, the prosthetic evolves to better serve the user.

11.2. Robotics

Robotic technology has also seen a significant rise in application within the field of bionic prosthetics and exoskeletons. Notably, robotic exoskeletons are making tremendous strides in giving individuals with mobility impairments newfound freedom.

These wearable devices can be used to assist individuals who have lost their walking abilities due to injury, illness, or age. They aid in restoring or augmenting the user's physical capabilities, enabling them to walk, run, and carry out many daily activities independently.

Robotic exoskeletons have been introduced into medical rehabilitation centers, providing patients with the necessary practice and assistance to regain their mobility. They are also making their way into industries where manual labor is required, helping to reduce worker fatigue and prevent injuries.

11.3. Advancements in Material Science: Lightness and Durability

Material science also plays a crucial role in the progression of this field. A significant challenge that researchers face is developing materials that are both lightweight and durable. Recent innovations have led to the creation of prosthetics from materials like titanium and carbon fiber.

These materials are light, ensuring that the prosthetic does not put unnecessary load on the user. At the same time, they are durable enough to withstand daily use without being prone to breakage or damage. Consequently, the user experiences more comfort and ease while using these devices.

11.4. Nanotechnology: The Smallest of Changes, The Largest of Impacts

Nanotechnology has the potential to bring considerable

advancements in the world of bionic prosthetics and exoskeletons. It can lead to the creation of "smart" prosthetics capable of mirroring the complexity of human tissues. Moreover, it can be used to enhance compatibility with the body and reduce the risk of rejection or inflammation.

One of the most exciting prospects of nanotechnology is the creation of nanosensors. These can be embedded in the prosthetic devices and can detect neural signals. They can help bridge the gap between the user's thoughts and the prosthetic device, leading to smoother device control.

11.5. Towards a More Simulated Experience: Haptic Feedback Systems

Introducing haptic feedback systems into prosthetics can change the game significantly. Haptic technology can provide feedback to the user about the feel, pressure, or movement they would naturally receive from their limb. This technology provides a more interactive experience, facilitating a closer connection between the user and their bionic limb, and ultimately leading to a more natural and intuitive use of the device.

11.6. Accessibility and Affordability

Prosthetics and exoskeletons need to be accessible and affordable for every individual who requires them. Therefore, efforts are being made to reduce the production cost without compromising the device's performance level.

Open-source development is a key player in this regard. Many researchers and developers are now open-sourcing their design and development processes. This sharing economy allows for rapid and less expensive prototyping, and it invites collaboration from all over the world.

The future of bionic limbs and exoskeletons flashes before us, promising a world where disability is not an end but a beginning of an extraordinary journey. Emerging trends and innovations are continually pushing the boundaries of what is possible, instigating a wave of transformation that is set to revolutionize human ability and potential in the years to come. We may not be able to cast a reassuring gaze upon the entire journey ahead, but the first few groundbreaking steps are becoming visible in the exciting landscape of prosthetics and exoskeletons.

www.ingramcontent.com/pod-product-compliance
Lightning Source LLC
Chambersburg PA
CBHW072219290526
45794CB00007B/2802